THE BATTERY THAT CHANGED EVERTHING

The Story of the Nuclear-Powered Smartphone

JOYCE KLINE

Copyright © 2024 by Joyce Kline.

All rights reserved.

Before this document is duplicated or reproduced in any manner, the publisher's consent must be gained. Therefore, the contents within can neither be stored electronically, transferred, nor kept in a database. Neither in Part nor full can the document be copied, scanned, faxed, or retailed without approval from the publisher or creator.

Table of Contents

Introduction 5
Unveiling the Power Within - The China Nuclear Battery 5
   The Evolution of Energy Storage 5
   Significance of Betavolt's Breakthrough 7

Chapter 1: Betavolt - Pioneering the Future 11
   Genesis of Betavolt 11
   Visionary Leadership and Scientific Innovation 13

Chapter 2: The Nuclear Battery Technology 18
   Understanding Nuclear Isotopes 18
   Energy Conversion with Diamond Semiconductors 21

Chapter 3: Betavolt's Initial Nuclear Battery 25
   Technical Specifications 25
   Power Output and Voltage Dynamics 28

Chapter 4: Safety and Reliability 32
   Layered Structure for Enhanced Safety 32
   Operating in Extreme Conditions 35

Chapter 5: Applications Beyond Smartphones 40
   Aerospace and AI Integration 40
   Medical Devices and Micro-Robots 44

Chapter 6: The Roadmap Ahead - Navigating Betavolt's Vision for 2025 and Beyond 48
   Betavolt's Plans for 2025 48

Scaling Up Production and Commercial Applications 52

Chapter 7: Comparative Analysis 58

Betavolt's Nuclear Battery vs. Conventional Technologies 58

Global Perspectives on Miniaturized Nuclear Batteries ... 62

Chapter 8: Addressing Concerns - Radiation and Environmental Impact 67

Betavolt's Assurance on Safety 67

Environmental Implications and Regulatory Considerations 71

Chapter 9: Impacts on Consumer Electronics 77

Revolutionizing Mobile Phones 77

Extending the Lifespan of Drones and Other Devices 82

Chapter 10: Future Prospects and Industry Landscape 87

The Potential Disruption in the Electronics Industry 87

Collaborations, Competitions, and Market Trends 91

Chapter 11: Ethical Considerations and Public Perception 97

Public Perception of Nuclear Energy in Consumer Devices 97

Ethical Implications and Societal Acceptance 101

Conclusion 106

Summing Up Betavolt's Contribution 106

Envisioning a Charged Future 110

Introduction

Unveiling the Power Within - The China Nuclear Battery

The Evolution of Energy Storage

In the ever-evolving landscape of technology, the quest for efficient and sustainable energy storage has been a driving force. From the humble beginnings of basic batteries to the sophistication of lithium-ion cells, the evolution has been relentless. The limitations of traditional energy storage methods, particularly in the context of portable electronics and emerging technologies, have spurred researchers to explore innovative solutions.

The narrative of energy storage has witnessed notable milestones, such as the development of rechargeable lithium-ion batteries, heralding a new era for portable electronic devices. However, as demands for longer-lasting power sources and environmental considerations loom large, the spotlight shifted to unconventional approaches.

Enter Betavolt, a Chinese startup that has propelled the discourse on energy storage into uncharted territories. Their groundbreaking work in the field of nuclear batteries signifies a paradigm shift, challenging the status quo and pushing the boundaries of what we perceive as feasible in the realm of power storage.

Significance of Betavolt's Breakthrough

The emergence of Betavolt as a key player in the energy storage domain is not merely a technological advancement; it is a watershed moment with far-reaching implications. Betavolt's breakthrough in developing a nuclear battery that claims to power smartphones for an impressive 50 years without charging has sent ripples across the scientific community and industry alike.

The significance of this breakthrough lies not only in its potential to revolutionize the consumer electronics market but also in addressing pressing global concerns. The world is at a crossroads where the demand for energy-efficient and sustainable technologies intersects with the imperative to reduce our reliance on conventional power sources with environmental ramifications.

Betavolt's nuclear battery is not just a technological marvel; it represents a bridge between the ambitions of technological progress and the imperatives of environmental sustainability. As the world grapples with the consequences of climate change and the need for cleaner energy alternatives, Betavolt's innovation takes center stage as a potential game-changer.

The societal impact of this breakthrough is profound. Imagine a world where smartphones, ubiquitous in modern life, no longer contribute to the growing e-waste problem due to their limited battery lifespans. Betavolt's technology has the potential to disrupt the very foundation of our interactions with electronic devices, fostering a more sustainable and eco-friendly future.

Furthermore, the geopolitical implications of China taking a lead in pioneering such transformative technologies cannot be overstated. In an era where technological prowess is intertwined with global influence, Betavolt's nuclear battery places China at the forefront of the race to define the future of energy storage.

In essence, the significance of Betavolt's breakthrough extends beyond the technological marvel itself; it reverberates through the corridors of environmental stewardship, societal evolution, and geopolitical dynamics. As we delve deeper into the intricacies of this revolutionary technology, we unravel not just the technical details but the potential to reshape the very fabric of our energy-dependent existence.

Chapter 1: Betavolt - Pioneering the Future

Genesis of Betavolt

The journey of Betavolt from a conceptual spark to a trailblazing entity in the energy storage landscape is marked by a compelling genesis. Founded against the backdrop of a rapidly evolving technology sector, Betavolt emerged as a response to the limitations posed by traditional batteries and the growing demand for sustainable energy solutions.

The genesis of Betavolt can be traced to a group of visionary scientists and entrepreneurs who saw the potential of harnessing nuclear energy for practical, everyday applications. The founders envisioned a future where energy storage transcends the boundaries of conventional methods and unlocks new possibilities for powering a wide array of devices with unprecedented longevity.

Betavolt's early years were characterized by intensive research and development, with a focus on understanding the intricacies of nuclear energy and its potential as a reliable power source. The company's commitment to pushing the boundaries of what was considered feasible in the field of energy storage laid the foundation for what would become a groundbreaking innovation.

The genesis of Betavolt is not just a story of technological ambition; it reflects a deep understanding of the pressing challenges in the energy sector. As the world grapples with the environmental impact of conventional energy sources, Betavolt emerged as a beacon of hope, offering a glimpse into a future where sustainable and long-lasting energy storage is not just a possibility but a reality.

Visionary Leadership and Scientific Innovation

At the heart of Betavolt's success lies visionary leadership that navigated uncharted territories with strategic acumen and unwavering commitment. The leadership team at Betavolt brought together a diverse group of experts, including physicists, engineers, and business leaders, each contributing a unique perspective to the ambitious goal of revolutionizing energy storage.

The visionary leadership at Betavolt fostered an environment that encouraged scientific innovation and exploration. The company became a melting pot of ideas, where researchers were given the freedom to dream big and pursue unconventional solutions. This culture of innovation propelled Betavolt beyond the realms of theoretical possibilities into the realm of tangible breakthroughs.

Scientific innovation at Betavolt took center stage as researchers delved into the complexities of nuclear energy and its conversion into a practical power source. The journey involved overcoming numerous technical challenges, from developing miniaturized nuclear modules to ensuring safety in the utilization of radioactive isotopes. The meticulous attention to detail and the relentless pursuit of excellence in scientific endeavors characterized Betavolt's approach.

The collaboration between visionary leadership and scientific innovation culminated in the development of the world's first miniaturized atomic energy system — the nuclear battery. This achievement stands as a testament to the power of interdisciplinary collaboration and the transformative potential of daring to explore the unknown.

Betavolt's leadership understood that true innovation requires not only technical prowess but also the ability to navigate regulatory landscapes, build strategic partnerships, and communicate the benefits of their breakthrough to the world. The visionary leaders orchestrated a holistic approach that positioned Betavolt not just as a technological innovator but as a key player in shaping the future of energy storage.

As we delve deeper into Betavolt's journey of pioneering the future, we unravel not only the technical intricacies of their innovation but also the human aspect — the visionaries who dared to dream and the scientists who turned those dreams into a reality that could reshape the world's relationship with energy.

Chapter 2: The Nuclear Battery Technology

Understanding Nuclear Isotopes

At the core of Betavolt's revolutionary nuclear battery technology lies a profound understanding of nuclear isotopes — the building blocks that harness the power of atomic energy for practical applications. Nuclear isotopes are variants of chemical elements with the same number of protons but different numbers of neutrons. It is this subtle difference that imparts unique properties to each isotope, making them ideal candidates for energy production.

Betavolt's nuclear battery utilizes 63 nuclear isotopes compactly packed into a module smaller than a coin. This dense arrangement is a testament to the meticulous engineering and scientific precision applied to harness the maximum energy potential from these isotopes. The selection of isotopes is a crucial aspect of Betavolt's technology, ensuring a delicate balance between energy output, stability, and safety.

The process begins with the controlled decay of these isotopes. As they undergo decay, they emit particles and release energy. It is this released energy that serves as the fundamental power source for the nuclear battery. The challenge lies in harnessing this energy efficiently and converting it into a usable form, a task at which Betavolt has excelled.

Betavolt's understanding of nuclear isotopes goes beyond mere utilization; it encompasses safety considerations as well. The selected isotopes undergo a transformation during the decay period, ultimately turning into stable, non-radioactive isotopes. This feature not only ensures the safety of the technology but also addresses concerns related to environmental impact and long-term usability.

Energy Conversion with Diamond Semiconductors

The second key pillar of Betavolt's nuclear battery technology involves the intricate process of energy conversion using diamond semiconductors. While the concept of using semiconductors in energy conversion is not new, Betavolt's innovation lies in the application of diamond semiconductors, a material known for its exceptional properties.

Diamond semiconductors offer a unique combination of high thermal conductivity, wide bandgap, and excellent electrical insulating properties. These characteristics make diamonds ideal for handling the challenges posed by the decay process of nuclear isotopes. The high thermal conductivity ensures efficient heat dissipation, while the wide bandgap allows for better control over the flow of electrons.

Betavolt's scientists developed a thin single-crystal diamond semiconductor, merely 10 microns thick, to facilitate the energy conversion process. This thin layer ensures minimal energy loss and optimal performance. Placed between two diamond semiconductor converters, a 2-micron-thick nickel-63 sheet serves as the bridge between the decaying isotopes and the generation of electrical current.

The conversion of decay energy into electrical current is a precise dance orchestrated by the unique properties of diamond semiconductors. As the radioactive isotopes emit particles during decay, these particles interact with the diamond semiconductors, causing the release of electrons. These liberated electrons form an electric current, which can then be harnessed for various applications.

The use of diamond semiconductors not only enhances the efficiency of the energy conversion process but also contributes to the overall safety of Betavolt's nuclear battery. Diamonds are renowned for their resilience and stability, ensuring that the semiconductors can withstand the conditions imposed by the decay process and operate reliably over an extended period.

In essence, Betavolt's nuclear battery technology represents a harmonious integration of nuclear isotopes and diamond semiconductors, each playing a pivotal role in a symphony of energy production. Understanding the nuances of these components unveils the complexity of Betavolt's achievement and paves the way for a deeper appreciation of the transformative potential embedded in this groundbreaking technology.

Chapter 3: Betavolt's Initial Nuclear Battery

Technical Specifications

Betavolt's foray into the realm of nuclear batteries is marked by the introduction of their initial marvel, the BV100. This miniature powerhouse defies conventional norms, representing a leap forward in energy storage technology. The technical specifications of the BV100 unveil a meticulous engineering feat that combines precision, innovation, and a vision for a sustainable future.

The BV100, in its compact form, delivers 100 microwatts of power, setting a new benchmark for the energy output achievable from a nuclear battery of its size. Measuring a mere 15x15x5 cubic millimeters, the BV100 is a testament to Betavolt's commitment to miniaturization without compromising efficiency.

One of the key technical aspects of the BV100 is its use of 63 nuclear isotopes compactly arranged within the battery module. This strategic placement ensures optimal utilization of the isotopes' energy potential, resulting in a consistent and reliable power supply. The selection of isotopes, their decay characteristics, and the overall architecture of the battery module contribute to the technical prowess of the BV100.

Safety features are ingrained in the technical design, with Betavolt employing a layered structure to prevent the battery from catching fire or exploding when subjected to sudden force. This layered approach not only enhances safety but also reflects Betavolt's commitment to producing technology that aligns with rigorous safety standards.

Looking beyond the present, Betavolt has laid out a roadmap for future iterations. The company envisions a battery with 1 watt of power by 2025, a significant leap from the initial 100 microwatts. This forward-looking approach underscores Betavolt's dedication to continuous improvement and scaling up the impact of their technology.

Power Output and Voltage Dynamics

The power output and voltage dynamics of the BV100 are pivotal aspects that define its usability and applicability in various scenarios. Delivering 100 microwatts of power, this nuclear battery may seem modest compared to traditional power sources, but its significance lies in its sustained output over an extended period.

The voltage dynamics of the BV100 are equally noteworthy, with a voltage of 3V. This voltage level aligns with the requirements of many low-power electronic devices, making the BV100 a viable candidate for a wide range of applications. The balance between power output and voltage dynamics is a delicate equilibrium achieved through meticulous engineering, ensuring compatibility with diverse electronic systems.

The small size of the BV100 allows for multiple units to be connected, creating the potential for scalable power solutions. Betavolt's vision extends beyond individual devices, envisioning a future where the modular nature of their nuclear batteries enables seamless integration into larger systems, amplifying the overall power output.

Betavolt's foresight extends to addressing the need for diversified power solutions in various industries. The BV100's technical specifications position it as a versatile power source for applications ranging from aerospace and AI equipment to medical devices, microprocessors, small drones, and micro-robots. This adaptability underscores the potential for Betavolt's nuclear battery to revolutionize multiple sectors simultaneously.

As we delve into the technical intricacies of Betavolt's initial nuclear battery, it becomes evident that the BV100 is not just a technological innovation but a catalyst for reshaping the landscape of energy storage. Its technical specifications serve as the foundation for a future where miniature yet powerful nuclear batteries become integral to powering the devices that define our daily lives. Betavolt's commitment to pushing the boundaries of what is achievable in energy storage sets the stage for a charged future where sustainability and efficiency coexist in harmony.

Chapter 4: Safety and Reliability

Layered Structure for Enhanced Safety

In the pursuit of groundbreaking nuclear battery technology, Betavolt places safety at the forefront of its design principles. The BV100, Betavolt's initial nuclear battery, boasts a layered structure that not only enhances its safety profile but also sets a new standard for secure and reliable energy storage solutions.

The layered structure serves as a protective shield, mitigating potential risks associated with the use of radioactive isotopes within the battery module. Betavolt's engineers meticulously designed this structure to prevent undesirable outcomes, such as the battery catching fire or exploding, particularly when subjected to sudden force or external stresses.

Each layer within the structure plays a crucial role in safeguarding the integrity of the nuclear battery. The outermost layer serves as a robust protective shell, acting as the first line of defense against external factors. This layer is engineered to resist physical impact and provide containment in the event of unforeseen circumstances.

Beneath the protective shell lies an insulating layer that acts as a thermal barrier. This insulating layer is designed to dissipate heat efficiently, ensuring that the nuclear battery remains within a safe operating temperature range. Efficient heat dissipation is critical in preventing overheating and maintaining the stability of the isotopes within the module.

The heart of the layered structure is the core containing the nuclear isotopes. This core is designed with redundancy and fail-safe mechanisms to prevent any potential breaches. The strategic arrangement of the isotopes within the core ensures uniform decay and energy release, contributing to the overall safety and reliability of the nuclear battery.

Betavolt's commitment to safety is not confined to theoretical considerations; it extends to real-world applications and scenarios. The layered structure is a testament to the company's dedication to providing consumers and industries with energy storage solutions that prioritize both performance and safety. As Betavolt continues to innovate, the layered safety architecture will likely remain a cornerstone of their designs, setting a benchmark for the industry.

Operating in Extreme Conditions

One of the hallmarks of Betavolt's nuclear battery technology is its ability to operate in extreme conditions, pushing the boundaries of where traditional batteries might falter. The BV100 is engineered to function flawlessly across a wide temperature range, from -60 degrees Celsius to 120 degrees Celsius. This adaptability positions Betavolt's nuclear battery as a robust and reliable power solution for diverse environments and applications.

Extreme temperatures pose a significant challenge to many conventional batteries, impacting their performance, lifespan, and, in some cases, safety. Betavolt's nuclear battery, however, thrives in conditions that would render other power sources ineffective. The ability to operate in extreme cold or hot environments opens up new possibilities for applications in aerospace, remote scientific stations, and various industrial settings.

The technology's resilience in extreme conditions is intricately tied to the properties of the diamond semiconductors used in the energy conversion process. Diamonds, known for their exceptional thermal conductivity and stability, contribute to the robustness of the nuclear battery. The precise engineering of the battery's components ensures that it can maintain optimal performance even when subjected to temperature fluctuations beyond the scope of conventional batteries.

The versatility to operate in extreme conditions positions Betavolt's nuclear battery as a game-changer in scenarios where conventional power sources fall short. For example, applications in aerospace, where temperature variations during space missions are inevitable, stand to benefit from the reliability and endurance of Betavolt's technology.

Moreover, the adaptability to extreme conditions aligns with Betavolt's vision of creating energy solutions for a wide array of applications, from powering medical devices in extreme environments to supporting the energy needs of AI equipment and drones operating in challenging conditions.

As Betavolt continues to refine and expand its nuclear battery technology, the capacity to operate in extreme conditions is likely to open doors to new possibilities across industries. The robustness and reliability of the BV100 lay the groundwork for a future where energy storage solutions can thrive in environments previously deemed inhospitable. Betavolt's commitment to safety and reliability, coupled with its innovative engineering, positions the company at the forefront of transformative developments in the energy storage landscape.

Chapter 5: Applications Beyond Smartphones

Aerospace and AI Integration

The impact of Betavolt's nuclear battery technology extends far beyond the realm of smartphones, ushering in a new era of possibilities in aerospace and artificial intelligence (AI) integration. The BV100's compact size, coupled with its impressive power output and reliability, positions it as a transformative power source for critical applications in these high-tech domains.

In the aerospace industry, where every ounce of weight and inch of space matters, the BV100's miniature form factor becomes a game-changer. Traditional power sources often struggle to meet the stringent weight and size constraints of aerospace applications, limiting the capabilities of satellites, space probes, and other space-bound devices. Betavolt's nuclear battery, with its small dimensions and scalable design, offers a solution to this longstanding challenge.

Satellites, for instance, require long-lasting, reliable power sources for sustained missions in space. The BV100's ability to deliver a continuous and stable power supply for extended periods aligns seamlessly with the demands of space exploration. Moreover, the adaptability to extreme temperature variations further enhances its suitability for the harsh conditions of outer space.

Beyond satellites, the integration of Betavolt's nuclear batteries into AI systems represents a paradigm shift in the field of artificial intelligence. AI applications, whether embedded in robotics or advanced computing systems, demand power sources that can provide consistent and efficient energy over extended durations. The BV100's nuclear battery, with its envisioned 1-watt power capability, becomes a compelling choice for fueling the next generation of AI-driven technologies.

The synergy between Betavolt's nuclear battery and AI systems lies in their shared vision for continuous, uninterrupted operation. In scenarios where traditional power sources may require frequent recharging or replacement, the BV100's longevity and reliability offer a significant advantage. This is particularly crucial in AI applications where uninterrupted processing and data analysis are paramount.

As AI continues to evolve and find applications in diverse industries, from autonomous vehicles to advanced robotics, the demand for compact, efficient, and long-lasting power sources will escalate. Betavolt's nuclear battery emerges as a frontrunner in meeting these demands, contributing to the realization of AI-driven technologies that operate seamlessly and independently.

Medical Devices and Micro-Robots

The marriage of Betavolt's nuclear battery technology with the field of medicine opens up a realm of possibilities for powering medical devices and micro-robots. In healthcare, where reliability and longevity are critical, the BV100's attributes become instrumental in reshaping the landscape of medical technology.

Medical devices, ranging from implantable devices like pacemakers to diagnostic equipment, often rely on batteries to function. Betavolt's nuclear battery, with its safety features, compact size, and extended lifespan, becomes an ideal candidate for powering these devices. The layered structure that enhances safety ensures that medical devices incorporating Betavolt's technology meet the stringent standards of the healthcare industry.

Implantable medical devices, such as pacemakers and cochlear implants, require power sources that can operate reliably within the human body. Betavolt's assurance of no external radiation, coupled with the transformation of isotopes into stable, non-radioactive forms post-decay, aligns with the safety requirements for devices within the human body. The prospect of a nuclear battery that can power these devices for extended periods without the need for frequent replacements offers a significant advancement in patient care.

Micro-robots, with applications in targeted drug delivery and minimally invasive surgeries, benefit from the BV100's compact design and energy efficiency. These tiny devices operate in intricate environments within the human body, necessitating power sources that can support their functions over time. Betavolt's nuclear battery provides a reliable and long-lasting energy solution for powering micro-robots, paving the way for innovations in medical treatments and procedures.

Furthermore, the adaptability of Betavolt's technology to extreme conditions enhances its potential applications in medical scenarios. From powering devices in emergency medical situations to supporting medical equipment in challenging environments, the BV100 becomes a versatile energy solution that transcends the limitations of traditional batteries.

As Betavolt's nuclear battery technology continues to evolve, its applications in aerospace, AI, and healthcare are poised to redefine industry standards. The potential to power critical systems reliably and sustainably positions Betavolt as a catalyst for advancements that go beyond the capabilities of existing power sources. The fusion of technological innovation and real-world applications heralds a future where Betavolt's nuclear battery becomes an indispensable component in shaping the trajectory of progress across diverse fields.

Chapter 6: The Roadmap Ahead - Navigating Betavolt's Vision for 2025 and Beyond

Betavolt's Plans for 2025

As Betavolt continues to redefine the landscape of energy storage with its pioneering nuclear battery technology, a glance into the future reveals an ambitious roadmap for the year 2025. The plans outlined by Betavolt offer insights into the company's strategic vision, highlighting key milestones, technological advancements, and the commitment to pushing the boundaries of what is achievable in the realm of sustainable energy.

At the heart of Betavolt's plans for 2025 is the aspiration to produce a nuclear battery with a power output of 1 watt. This targeted enhancement in power represents a significant leap from the initial 100 microwatts delivered by the BV100. The pursuit of higher power output is driven by the company's recognition of the evolving demands of modern technology and the need for more robust energy solutions.

The transition from microwatts to watts in power output signifies Betavolt's commitment to providing energy solutions that align with the increasing power requirements of advanced electronic devices. From aerospace applications to AI systems with growing computational complexities, the 1-watt nuclear battery aims to empower a diverse range of technologies with sustained and efficient energy.

The journey towards achieving this milestone involves a confluence of scientific innovation, engineering excellence, and a dedication to continuous improvement. Betavolt's team of scientists and engineers are poised to explore new materials, refine manufacturing processes, and optimize the design of nuclear battery modules to realize the envisioned 1-watt power output.

Beyond the technical aspects, Betavolt's plans for 2025 extend to forging strategic partnerships and collaborations.

Collaborative efforts with research institutions, technology companies, and industry leaders are anticipated to accelerate the development and adoption of Betavolt's nuclear battery technology. The synergy between Betavolt and its partners is envisioned to amplify the impact of nuclear batteries across diverse sectors.

Furthermore, Betavolt aims to deepen its engagement with regulatory bodies and environmental organizations to ensure that its nuclear batteries align with global safety and sustainability standards. Transparent communication about the safety features, environmental impact, and potential applications of the technology is integral to

Betavolt's approach as it navigates the pathway to 2025 and beyond.

Scaling Up Production and Commercial Applications

Scaling up production is a pivotal pillar of Betavolt's roadmap, with a focus on transitioning from pilot testing to mass production of its nuclear batteries. The success of the BV100 in pilot testing has provided valuable insights and validation of Betavolt's technology, laying the foundation for the next phase of commercialization.

Betavolt envisions a future where its nuclear batteries are not just technological marvels but integral components in a wide array of commercial applications. The scalability of production processes is crucial in realizing this vision, ensuring that Betavolt's energy solutions can meet the demands of industries ranging from consumer electronics to healthcare and beyond.

Efficient and scalable manufacturing processes are central to Betavolt's strategy for making nuclear batteries accessible to a broader market. Streamlining production, optimizing resource utilization, and leveraging economies of scale are key elements in ensuring that the benefits of Betavolt's technology reach consumers, businesses, and industries worldwide.

Commercial applications of Betavolt's nuclear batteries extend across various sectors, each presenting unique challenges and opportunities. The adaptability of the technology to diverse scenarios, from powering smartphones to supporting critical systems in aerospace, positions Betavolt as a versatile player in the energy storage market.

In consumer electronics, the prospect of smartphones that never need to be charged is a paradigm shift that resonates with the daily lives of individuals. Betavolt's nuclear batteries, with their extended lifespan and reliability, redefine the user experience, eliminating the inconvenience of frequent charging cycles and enhancing the overall usability of electronic devices.

The scalability of Betavolt's technology also holds promise for applications in the automotive industry. Electric vehicles, with their increasing prevalence, stand to benefit from the integration of nuclear batteries that offer prolonged driving ranges and reduced dependence on charging infrastructure. Betavolt's roadmap includes exploring partnerships with automakers to drive innovation in sustainable transportation.

In the realm of healthcare, the potential applications of Betavolt's nuclear batteries in powering medical devices and implants are revolutionary. The reliability and safety features of the technology align with the stringent requirements of the healthcare industry, presenting opportunities to enhance patient care and advance medical treatments.

Betavolt's roadmap for scaling up production and commercial applications envisions a future where nuclear batteries become commonplace in powering devices, systems, and technologies that shape our daily lives. The company's commitment to advancing sustainable energy solutions positions it at the forefront of transformative developments in the energy storage landscape.

As Betavolt navigates the roadmap ahead, the convergence of technological innovation, strategic partnerships, and a commitment to sustainability propels the company toward a future where nuclear batteries play a central role in powering a world hungry for efficient, reliable, and environmentally conscious energy solutions.

Chapter 7: Comparative Analysis

Betavolt's Nuclear Battery vs. Conventional Technologies

Betavolt's foray into the realm of energy storage with its groundbreaking nuclear battery has sparked a keen interest in how it compares with conventional technologies. A comparative analysis sheds light on the strengths, limitations, and unique attributes that distinguish Betavolt's nuclear battery from traditional energy storage solutions.

One of the key differentiators is the energy density offered by Betavolt's nuclear battery. Traditional lithium-ion batteries, which dominate the consumer electronics market, have served as reliable power sources. However, their energy density is limited, necessitating frequent recharging. Betavolt's nuclear battery, on the other hand, boasts an impressive energy density, enabling prolonged usage without the need for frequent charging cycles.

The compact size of Betavolt's nuclear battery further sets it apart from conventional technologies. Lithium-ion batteries, despite their widespread use, face challenges related to size and weight. The miniaturized form factor of Betavolt's nuclear battery opens avenues for applications in aerospace, medical devices, and other scenarios where space is a premium. This compact design aligns with the increasing trend toward smaller, more efficient electronic devices.

Moreover, Betavolt's nuclear battery promises an extended lifespan without degradation in capacity—a critical aspect where conventional batteries often falter. The continuous decay of isotopes within Betavolt's nuclear battery ensures a consistent power output over the years, addressing one of the persistent challenges faced by traditional batteries—their diminishing capacity over time.

Safety considerations also differentiate Betavolt's nuclear battery from conventional technologies. The layered structure of the nuclear battery is designed to prevent fire or explosion when subjected to external force, a concern often associated with lithium-ion batteries. Betavolt's commitment to safety extends to its claim that the battery can operate in a wide temperature range without compromising performance, a feature that enhances its suitability for diverse environments.

While Betavolt's nuclear battery demonstrates several advantages, it is essential to acknowledge areas where conventional technologies currently maintain their edge. The maturity and established infrastructure of lithium-ion batteries contribute to their widespread adoption and accessibility. Charging infrastructure, compatibility with existing devices, and familiarity with lithium-ion technology remain factors that Betavolt's nuclear battery will need to navigate as it seeks broader acceptance.

Global Perspectives on Miniaturized Nuclear Batteries

The emergence of miniaturized nuclear batteries, epitomized by Betavolt's technological breakthrough, has garnered attention on a global scale. Exploring global perspectives provides insights into the reception, challenges, and potential applications of miniaturized nuclear batteries beyond Betavolt's endeavors.

China, the birthplace of Betavolt's nuclear battery, stands at the forefront of embracing this transformative technology. The strategic integration of miniaturized nuclear batteries aligns with China's commitment to technological innovation and sustainable energy solutions. The government's support, coupled with initiatives under the 14th Five-Year Plan, reflects a concerted effort to position China as a leader in the field of advanced energy storage.

In the United States and Europe, research institutions and companies are also actively engaged in the development of miniaturized nuclear batteries. The historical exploration of this technology for space missions and remote scientific stations has laid the groundwork for contemporary advancements. Efforts to miniaturize and commercialize nuclear batteries align with a shared vision of enhancing energy efficiency and sustainability.

While Betavolt's nuclear battery has garnered attention for its potential to revolutionize electronics, it is part of a broader global narrative seeking solutions to the limitations of conventional batteries. The pursuit of alternatives to lithium-ion batteries is not limited to China; it resonates across borders as nations recognize the need for energy storage innovations to meet the demands of an evolving technological landscape.

Global perspectives also illuminate the regulatory considerations and public perceptions surrounding miniaturized nuclear batteries. As with any revolutionary technology, questions related to safety, environmental impact, and disposal mechanisms are paramount. International collaboration and standardization efforts are crucial in addressing these concerns and establishing a framework for the responsible development and deployment of miniaturized nuclear batteries.

The comparative analysis of Betavolt's nuclear battery with conventional technologies offers a glimpse into the evolving dynamics of the energy storage market. While Betavolt's technology introduces game-changing features, it is situated within a global context where multiple players contribute to the collective pursuit of sustainable and efficient energy solutions.

As miniaturized nuclear batteries continue to capture global attention, the landscape is marked by collaboration, competition, and shared aspirations for a future where energy storage transcends the limitations of the past. Betavolt's journey is a testament to the transformative potential of nuclear battery technology, contributing to a broader conversation on reshaping the future of energy storage on a global scale.

Chapter 8: Addressing Concerns - Radiation and Environmental Impact

Betavolt's Assurance on Safety

The unveiling of Betavolt's revolutionary nuclear battery has inevitably raised concerns related to radiation safety. As with any technology harnessing nuclear elements, the potential risks associated with radiation exposure demand careful consideration. Betavolt, cognizant of these concerns, has undertaken rigorous measures to address and mitigate potential risks, offering assurances on the safety of its nuclear batteries.

One of the primary concerns associated with nuclear energy is external radiation emission.

Betavolt emphatically asserts that its nuclear batteries are designed with safety as a paramount consideration. The layered structure of the battery serves as a protective barrier, preventing external radiation leakage. This design not only safeguards users but also positions Betavolt's nuclear batteries for applications where safety is a critical factor, such as in medical devices implanted within the human body.

To provide further reassurance, Betavolt emphasizes the choice of nickel-63 as the radioactive element in its nuclear battery. Nickel-63 is a beta-emitting isotope, emitting electrons as part of its decay process. This type of radiation is relatively low-energy and can be effectively contained within the layered structure of the battery. Betavolt contends that the controlled decay of nickel-63 ensures a stable and predictable energy release, minimizing potential hazards associated with high-energy radiation sources.

Betavolt's commitment to safety extends beyond the design of the nuclear battery to its intended applications. The company asserts that the BV100 battery is suitable for use in medical devices like pacemakers and cochlear implants. The prospect of a nuclear battery powering critical medical devices underscores Betavolt's confidence in the safety profile of its technology. The ability to function within the human body, coupled with the absence of external radiation, positions Betavolt's nuclear battery as a reliable and secure energy source for life-saving medical applications.

Another facet of Betavolt's safety assurance lies in the transformation of nickel-63 isotopes. After the decay period, the isotopes transform into a stable, non-radioactive isotope of copper. This transformation eliminates any environmental threat or pollution, assuaging concerns about the long-term impact of the nuclear battery on ecosystems. Betavolt's approach aligns with principles of environmental responsibility, ensuring that the deployment of its technology does not leave a lasting footprint of radioactive materials.

While Betavolt's assurances on safety are explicit, the broader acceptance of nuclear batteries, including those from Betavolt, hinges on transparent communication and collaboration with regulatory bodies. The dissemination of accurate information about safety features, radiation containment mechanisms, and real-world performance is integral to building trust among users, industry stakeholders, and regulatory authorities.

Environmental Implications and Regulatory Considerations

The environmental implications of deploying nuclear batteries extend beyond the safety of users to considerations related to waste management and ecological impact. Betavolt recognizes the importance of addressing these concerns proactively, aligning its approach with environmental responsibility and compliance with regulatory standards.

One notable aspect is the minimal environmental impact during the operational phase of Betavolt's nuclear batteries. The controlled decay of nickel-63, coupled with the subsequent transformation into a stable copper isotope, eliminates the emission of harmful pollutants during the energy conversion process. This characteristic sets Betavolt's nuclear batteries apart from conventional energy storage technologies that may rely on materials with significant environmental consequences.

The decommissioning phase of nuclear batteries is a critical juncture where environmental considerations come to the forefront. Betavolt, cognizant of the need for responsible waste management, outlines its commitment to ensuring that the disposal of nuclear batteries adheres to established regulations and guidelines. The transformation of isotopes into non-radioactive forms contributes to the eco-friendly disposition of the nuclear battery at the end of its lifecycle.

Regulatory considerations play a pivotal role in shaping the acceptance and deployment of nuclear batteries. Betavolt's engagement with regulatory bodies involves transparent communication about the safety features, operational characteristics, and environmental impact of its technology. Collaborative efforts with regulators contribute to the development of standards that govern the use of miniaturized nuclear batteries, ensuring that industry practices align with global safety and environmental norms.

Internationally, the regulatory landscape for nuclear technologies varies, necessitating a nuanced approach in navigating diverse jurisdictions. Betavolt's commitment to meeting and exceeding regulatory standards underscores its dedication to responsible innovation. Collaborative dialogues with regulatory authorities contribute to the establishment of a regulatory framework that fosters the safe and sustainable integration of nuclear batteries into diverse applications.

Public perception also plays a crucial role in the regulatory landscape. Betavolt's transparency in addressing concerns related to radiation, safety, and environmental impact is instrumental in shaping a positive perception of nuclear battery technology. Education and awareness initiatives contribute to informed decision-making among users, policymakers, and the general public, fostering a conducive environment for the responsible adoption of Betavolt's nuclear batteries.

In conclusion, Betavolt's approach to addressing concerns related to radiation and environmental impact reflects a dual commitment—to the safety of users and the preservation of the environment. The company's emphasis on safety features, responsible waste management, and collaboration with regulatory authorities positions Betavolt as a key player in the evolution of nuclear battery technology, contributing to a future where sustainable and safe energy storage solutions play a central role in powering the world.

Chapter 9: Impacts on Consumer Electronics

Revolutionizing Mobile Phones

The advent of Betavolt's nuclear battery technology heralds a new era in consumer electronics, with a particular focus on revolutionizing the ubiquitous mobile phone. The implications for mobile phones are profound, promising to reshape the way we perceive and interact with these essential devices.

One of the primary challenges faced by smartphone users is the need for frequent charging. Betavolt's nuclear battery addresses this challenge head-on by offering a power solution that boasts an extraordinary lifespan. The prospect of mobile phones powered by nuclear batteries introduces the concept of devices that no longer require daily or even weekly charging. Users can envision a future where the anxiety of running out of battery during a crucial moment becomes a thing of the past.

The extended lifespan of Betavolt-powered mobile phones is a game-changer for users who rely heavily on their devices for communication, productivity, and entertainment. Imagine a smartphone that can operate continuously for years without degradation in performance or the need for battery replacements. This transformative shift not only enhances the user experience but also contributes to reducing electronic waste associated with disposable batteries.

The compact size of Betavolt's nuclear batteries further complements the trend towards sleek and lightweight mobile devices. As smartphones continue to evolve in design and functionality, the miniaturized form factor of Betavolt's technology aligns seamlessly with the quest for thinner and more portable devices. The integration of nuclear batteries into mobile phones opens up possibilities for innovative designs and form factors that go beyond the limitations of traditional battery technologies.

Beyond the convenience of prolonged battery life, Betavolt's nuclear batteries introduce a level of reliability and consistency previously unseen in consumer electronics. Mobile phones often suffer from diminishing battery capacity over time, leading to reduced performance and the eventual need for battery replacements. Betavolt's technology, with its stable power output and extended lifespan, mitigates these concerns, offering users a reliable and enduring power source.

The impact on consumer behavior is noteworthy as well. The reliance on power banks, charging cables, and the perpetual search for electrical outlets becomes a thing of the past. Users can enjoy a sense of freedom, untethered from the constant need for recharging. This shift in behavior contributes to a more sustainable and user-centric approach to mobile phone usage, aligning with the evolving expectations of today's tech-savvy consumers.

While the revolutionizing impact on mobile phones is evident, Betavolt's nuclear battery technology extends its transformative influence to a broader spectrum of consumer electronics. Drones, in particular, stand out as a category poised to benefit significantly from this breakthrough.

Extending the Lifespan of Drones and Other Devices

Drones have emerged as versatile tools with applications ranging from recreational photography to industrial inspections. However, their potential is often limited by the constraints of battery life. Betavolt's nuclear battery technology promises to extend the lifespan of drones and other devices, unlocking new possibilities and applications.

One of the primary challenges faced by drone enthusiasts and professionals alike is the limited flight time offered by conventional batteries. Betavolt's nuclear battery, with its impressive power output and extended lifespan, addresses this challenge by enabling drones to operate for significantly longer durations. This breakthrough has far-reaching implications for industries such as agriculture, surveying, and surveillance, where prolonged aerial operations are essential.

The use of nuclear batteries in drones introduces a paradigm shift in how these devices are deployed. Imagine agricultural drones that can monitor crops continuously without the need for frequent recharging, or surveillance drones that can maintain extended patrols for enhanced security. The extended lifespan of Betavolt-powered batteries in drones opens up possibilities for more ambitious and sustained missions, revolutionizing the capabilities of these aerial devices.

Moreover, the safety and reliability features of Betavolt's nuclear batteries are crucial in the context of drones. The layered structure that prevents fire or explosion, coupled with the ability to operate in a wide temperature range, ensures that drones equipped with Betavolt's technology can perform reliably in diverse environmental conditions. This reliability is particularly significant in scenarios where the failure of a drone can have critical consequences.

Beyond drones, the impact of Betavolt's nuclear batteries extends to a myriad of other consumer electronics. Wearable devices, for instance, stand to benefit from the prolonged battery life and compact form factor offered by nuclear batteries. The prospect of smartwatches, fitness trackers, and health monitoring devices operating continuously for extended periods aligns with the evolving expectations of users who seek seamless integration of technology into their daily lives.

The integration of Betavolt's technology into various consumer electronics also has implications for sustainability. The longer lifespan of nuclear batteries reduces the frequency of replacements, contributing to a reduction in electronic waste. This aligns with global efforts to promote environmentally friendly practices in the design and use of electronic devices.

In conclusion, Betavolt's nuclear battery technology not only revolutionizes mobile phones but also extends its transformative impact to a diverse array of consumer electronics. From drones with extended flight times to wearables with prolonged operational capabilities, the integration of nuclear batteries opens up a future where electronic devices operate more efficiently, sustainably, and reliably. As industries and consumers embrace this technological leap, the potential for innovation and new applications in consumer electronics becomes boundless.

Chapter 10: Future Prospects and Industry Landscape

The Potential Disruption in the Electronics Industry

Betavolt's groundbreaking nuclear battery technology not only represents a milestone in energy storage but also holds the potential to disrupt the entire electronics industry. The implications of this innovation extend beyond incremental improvements, offering a transformative shift in how electronic devices are powered. As we delve into the future prospects and industry landscape, it becomes evident that Betavolt's nuclear batteries are poised to redefine the norms and dynamics of the electronics sector.

One of the key aspects of Betavolt's nuclear battery technology that promises disruption is the elimination of the perpetual need for recharging. The conventional paradigm of charging cycles, battery degradation, and the constant quest for power sources is challenged by the enduring nature of nuclear batteries. This paradigm shift has far-reaching consequences for consumer behavior, device design, and the entire ecosystem of products and services associated with electronic devices.

The consumer electronics market, dominated by the smartphone industry, is likely to witness a significant shake-up. The concept of smartphones that do not require daily charging introduces a new dimension of convenience for users. This shift in user behavior necessitates a reevaluation of existing business models centered around accessories like power banks, charging cables, and related peripherals. Companies in the electronics industry will need to adapt to a landscape where the emphasis shifts from addressing battery limitations to enhancing overall device capabilities.

Moreover, the potential disruption extends to the electric vehicle (EV) sector. Betavolt's nuclear batteries, with their compact size, high power output, and extended lifespan, present an intriguing alternative to conventional lithium-ion batteries. The implications for electric cars, drones, and other electric-powered vehicles are substantial. The prospect of vehicles that can operate for extended distances without frequent recharging challenges the current infrastructure and business models associated with EVs.

In the aerospace industry, the disruption is equally profound. Drones, satellites, and other aerospace applications stand to benefit from the prolonged operational capabilities offered by Betavolt's nuclear batteries. The prospect of satellites with extended missions, drones for surveillance and exploration with significantly longer flight times, and the overall enhancement of aerospace technologies redefine the possibilities in this sector.

Collaborations, Competitions, and Market Trends

As Betavolt's nuclear battery technology gains traction, the industry landscape is marked by a dynamic interplay of collaborations, competitions, and emerging market trends. The collaboration between Betavolt and various stakeholders, including tech companies, research institutions, and regulatory bodies, becomes pivotal in shaping the trajectory of this technology.

Collaborations with leading tech companies open avenues for the integration of nuclear batteries into a diverse range of products. Smartphone manufacturers, for instance, may explore partnerships with Betavolt to incorporate this revolutionary energy storage solution into their devices. Such collaborations not only accelerate the adoption of nuclear batteries but also position Betavolt as a key player in shaping the future of electronics.

Research institutions play a crucial role in advancing the technology itself. Collaborative research initiatives contribute to refining nuclear battery designs, addressing potential challenges, and exploring new applications. The synergy between Betavolt and research institutions fosters an environment of continuous innovation, pushing the boundaries of what is achievable with nuclear battery technology.

The competitive landscape in the electronics industry is inevitably influenced by the emergence of Betavolt's technology. Traditional battery manufacturers face the challenge of adapting to a landscape where the rules of the game are changing. Companies invested in lithium-ion battery technologies, for example, may find themselves in competition with the disruptive potential of nuclear batteries.

Startups and new entrants also play a role in shaping the competitive dynamics. As the industry witnesses a paradigm shift, nimble and innovative startups may seize opportunities to offer specialized solutions or niche applications that leverage the benefits of nuclear batteries. The competitive landscape becomes a dynamic arena where adaptability and innovation are key determinants of success.

Market trends in the wake of Betavolt's nuclear battery introduction are characterized by a growing emphasis on sustainability and reliability. Consumers are likely to gravitate towards products that align with eco-friendly practices, and the extended lifespan of nuclear batteries positions them as a sustainable energy storage solution. Reliability, both in terms of power output and safety features, becomes a defining factor in consumer preferences.

Regulatory trends also come into play as the industry navigates the introduction of a new and potentially disruptive technology. Collaborative efforts between Betavolt and regulatory bodies contribute to the establishment of standards and guidelines that ensure the safe and responsible integration of nuclear batteries into electronic devices. Regulatory frameworks evolve to accommodate the unique characteristics and considerations associated with this novel energy storage solution.

In conclusion, the future prospects and industry landscape shaped by Betavolt's nuclear battery technology are marked by the potential for substantial disruption. The elimination of charging constraints, collaborations with industry stakeholders, competitive dynamics, and emerging market trends collectively contribute to a landscape where the traditional norms of the electronics industry are challenged and redefined. As the technology continues to mature, its impact on consumer behavior, product design, and the overall trajectory of the electronics sector will become increasingly pronounced.

Chapter 11: Ethical Considerations and Public Perception

Public Perception of Nuclear Energy in Consumer Devices

The integration of nuclear energy into consumer devices, as exemplified by Betavolt's groundbreaking nuclear battery technology, raises intriguing questions about public perception and the broader ethical considerations surrounding the use of nuclear power in everyday electronics. As we delve into this realm, it becomes evident that the public's perception of nuclear energy plays a pivotal role in shaping the ethical landscape and societal acceptance of this transformative technology.

Historically, public perception of nuclear energy has been influenced by a complex interplay of factors, including historical events, media representation, and cultural attitudes towards nuclear technology. The mention of nuclear energy often evokes images of power plants, radioactive waste, and high-profile incidents such as Chernobyl and Fukushima. As a result, the initial reaction to the idea of nuclear batteries powering consumer devices may be met with skepticism and concerns rooted in pre-existing notions of nuclear energy.

However, Betavolt's nuclear battery technology introduces a nuanced shift in the narrative. The compact size, safety features, and the promise of extended device lifespans contribute to a more favorable context for public acceptance. The juxtaposition of nuclear energy with daily gadgets like smartphones and drones challenges traditional associations, paving the way for a reevaluation of nuclear power's role in our lives.

To understand public perception, it is crucial to engage in transparent and educational communication about the technology. The communication strategy should emphasize not only the benefits—such as extended battery life and reduced electronic waste—but also address potential concerns head-on. Clear and accessible information about the safety measures in place, the absence of external radiation, and the environmental impact of the nuclear batteries is essential in shaping a positive public perception.

Furthermore, the narrative around nuclear energy in consumer devices can draw parallels with other ubiquitous technologies that were initially met with skepticism. The introduction of electricity, for example, was once viewed with caution and skepticism. Over time, as the benefits and safety measures became evident, electricity became an integral part of modern life. Framing nuclear batteries within the context of technological evolution and progress can help reshape public perception and foster a more open-minded approach.

Ethical Implications and Societal Acceptance

The ethical considerations surrounding the use of nuclear energy in consumer devices extend beyond individual perceptions to broader societal implications. As Betavolt's nuclear battery technology emerges as a potential game-changer, ethical scrutiny becomes paramount in assessing its impact on individuals, communities, and the environment.

One of the primary ethical considerations lies in the production and disposal of nuclear batteries. The extraction and processing of radioactive materials, such as nickel-63, raise questions about the environmental and social impact of these processes. Ensuring responsible and sustainable sourcing practices, coupled with effective recycling methods for decommissioned batteries, becomes a key ethical imperative. Betavolt and similar innovators must be transparent about their supply chain practices, emphasizing a commitment to minimizing environmental harm and supporting ethical resource extraction.

Societal acceptance of nuclear batteries also hinges on equitable access and affordability. As this technology has the potential to revolutionize consumer electronics, ensuring that the benefits are accessible across diverse socioeconomic strata becomes an ethical imperative. Betavolt's commitment to scalable production and collaboration with various stakeholders, including regulatory bodies and industry partners, can contribute to the ethical distribution of this transformative technology.

Safety considerations are at the forefront of ethical discussions surrounding nuclear energy. Betavolt's layered structure, designed to prevent fire or explosion, and the assurance of no external radiation are crucial ethical safeguards. However, ongoing research and vigilance are essential to address any unforeseen safety concerns that may arise as the technology scales. Public trust in the ethical use of nuclear batteries hinges on continuous efforts to uphold safety standards and transparency in addressing potential risks.

The potential for geopolitical implications also adds a layer of ethical complexity. As China takes a pioneering role in developing and commercializing nuclear battery technology, questions arise about the geopolitical power dynamics in the electronics industry. Collaborative efforts, international regulatory frameworks, and ethical business practices are imperative to navigate these complexities and ensure that the global deployment of nuclear batteries aligns with ethical standards.

In conclusion, ethical considerations and societal acceptance play a pivotal role in the successful integration of nuclear energy into consumer devices. Betavolt's nuclear battery technology, while offering transformative benefits, must navigate the ethical landscape with transparency, responsibility, and a commitment to addressing potential concerns. By actively engaging with the public, adhering to ethical production and disposal practices, and fostering equitable access, Betavolt can contribute to reshaping perceptions and fostering societal acceptance of this innovative energy solution. As the journey unfolds, the ethical compass guiding nuclear battery technology will be essential in shaping a future where the benefits are maximized responsibly and ethically.

Conclusion

Summing Up Betavolt's Contribution

In the realm of energy storage and consumer electronics, Betavolt's groundbreaking nuclear battery technology stands as a testament to innovation that transcends boundaries. As we delve into the conclusion of this exploration, it becomes evident that Betavolt has not merely introduced a novel power source but has ushered in a paradigm shift with far-reaching implications for the future of technology and energy.

Summing up Betavolt's contribution requires acknowledging the journey from conceptualization to realization. The genesis of Betavolt as a Chinese startup dedicated to pushing the boundaries of energy storage is a story of visionary leadership and scientific ingenuity. The audacious goal of developing a nuclear battery that could power smartphones for 50 years without charging was met with skepticism and excitement in equal measure. Betavolt, under visionary leadership, embarked on a journey that culminated in the creation of the BV100 nuclear battery—an embodiment of their commitment to redefining the possibilities of energy storage.

The nuclear battery technology developed by Betavolt represents more than just a technical achievement. It encapsulates the spirit of pushing the limits of what is deemed possible. The intricate dance between nuclear isotopes, diamond semiconductors, and a commitment to safety culminated in a compact energy solution that challenges the status quo. The technical specifications, with the BV100 delivering 100 microwatts of power and the promise of scaling up to 1 watt by 2025, underscore Betavolt's trajectory toward achieving unprecedented milestones.

Envisioning a charged future requires looking beyond the immediate applications of nuclear batteries. Betavolt's vision extends to a world where mobile phones never need to be charged, drones can fly indefinitely, and electronic devices operate seamlessly without the constraints of traditional battery limitations. This future is not just a fanciful notion but a tangible prospect as Betavolt initiates pilot testing and lays the groundwork for mass production.

The safety and reliability features embedded in Betavolt's nuclear battery are pivotal in shaping this charged future. The layered structure that prevents fire or explosion, the ability to operate in extreme temperatures, and the assurance of no external radiation contribute to a narrative of a technology that not only redefines convenience but also prioritizes safety. As Betavolt paves the way for applications beyond smartphones, including aerospace integration, medical devices, and micro-robots, the potential impact on various industries becomes increasingly profound.

Envisioning a Charged Future

As we envision a charged future catalyzed by Betavolt's nuclear battery technology, the implications ripple across industries and consumer experiences. Mobile phones, often tethered to charging cables and power banks, could become emblematic of a bygone era. The prospect of devices that operate continuously without the need for regular charging transforms the way we interact with technology.

In the consumer electronics landscape, the charged future heralds a new era of sustainability and reliability. The traditional lifecycle of electronic devices, marked by the constant need for charging and the eventual degradation of battery capacity, undergoes a fundamental shift. Devices powered by nuclear batteries operate with extended lifespans, reducing electronic waste and contributing to a more sustainable approach to technology consumption.

The charged future extends beyond the realm of personal devices. In aerospace and AI integration, the potential for indefinite operation without the constraints of conventional power sources opens avenues for exploration and innovation. Drones equipped with nuclear batteries could redefine surveillance, exploration, and scientific research by operating continuously for extended durations. The vision of aerospace technologies that are not bound by the limitations of traditional power sources opens doors to possibilities that were once deemed impractical.

Medical devices, micro-robots, and advanced sensors, powered by Betavolt's nuclear batteries, promise enhanced capabilities and reliability. The prospect of medical implants that operate for decades without the need for replacement transforms healthcare paradigms. Micro-robots designed for intricate tasks, powered by a continuous and reliable energy source, become instrumental in fields ranging from manufacturing to healthcare.

Betavolt's Plans for 2025 and Scaling Up Production

Looking forward, Betavolt's plans for 2025 provide a glimpse into the company's trajectory and the potential evolution of nuclear battery technology. The commitment to produce a battery with 1 watt of power by 2025 speaks to Betavolt's dedication to continuous improvement and scaling up production. The small size of these batteries, measuring 15x15x5 cubic millimeters, allows for multiple units to be connected, presenting possibilities for increased power output and diverse applications.

Scaling up production is a critical facet of Betavolt's roadmap. As the technology advances from pilot testing to mass production, the impact on the consumer electronics market becomes increasingly tangible. The scalability of nuclear battery production has the potential to reshape the dynamics of supply chains, manufacturing processes, and collaborations within the electronics industry.

Collaborations with stakeholders, including tech companies, research institutions, and regulatory bodies, become instrumental in Betavolt's journey toward scaling up production. The collaborative efforts contribute not only to the refinement of the technology but also to the establishment of standards and guidelines that ensure the responsible integration of nuclear batteries into various applications.

In the charged future envisioned by Betavolt, the narrative extends beyond technical specifications and industrial applications. It becomes a story of collaboration, innovation, and the relentless pursuit of a future where the constraints of traditional energy storage are transcended. As Betavolt's plans for 2025 come to fruition, the impact on the electronics industry and beyond will be a testament to the transformative power of nuclear battery technology.

In conclusion, Betavolt's contribution to the landscape of energy storage and consumer electronics is more than a technological breakthrough—it is a narrative of redefining possibilities. The charged future envisioned by Betavolt extends beyond individual devices to encompass a societal shift towards sustainability, reliability, and a reimagined relationship with technology. As Betavolt's nuclear battery technology continues to evolve, its impact on industries, consumer experiences, and the broader trajectory of technological progress will undoubtedly leave an indelible mark on the charged future we collectively envision.

Made in United States
Troutdale, OR
03/09/2024